Lutz Peschke (ed.)

Visuality of the Anthropocene. A Concept for a Poster Exhibition

Lutz Peschke (ed.)

Visuality of the Anthropocene.
A Concept for a Poster
Exhibition

Bibliografische Information der Deutschen Nationalbibliothek:
Die Deutsche Nationalbibliothek verzeichnet diese Publikation in der
Deutschen Nationalbibliografie; detaillerte bibliografische Daten sind im
Internet über http://dnb.dnb.de abrufbar.

July 2019

© 2019 Lutz Peschke
Herstellung und Verlag:
BoD-Books on Demand, Norderstedt

ISBN: 978-3-7357-2540-0

Index

Lutz Peschke
The Impacts Of Greta Thunberg On Environmental
Communication In Germany.. 1

Seldağ Güneş Peschke
Environmental Law Clinics: Towards a New Era in
Legal Education 9

M. Mert Örsler
Geothermie Is Super Hot And As Beautiful As Ever 14

**Melina Diener, Marleen C. Schwalm, Manfred Schmidt,
Ruben Düchting**
Scheme Geothermal Power... 16

**Melina Diener, Marleen C. Schwalm, Manfred Schmidt,
Ruben Düchting**
Scheme Windpower.. 17

Simge Sadak
Pros And Cons Of Hydropower... 18

**Melina Diener, Marleen C. Schwalm, Manfred Schmidt,
Ruben Düchting**
Scheme Hydropower... 20

**Melina Diener, Marleen C. Schwalm, Manfred Schmidt,
Ruben Düchting**
Schemes Energy From Sunlights....................................... 21

**Melina Diener, Marleen C. Schwalm, Manfred Schmidt,
Ruben Düchting**
Scheme Biomass.. 22

M. Mert Örsler
Nuclear Power Is Not Wanted In Turkey............................. 23

Öykü Öncül
The Long Life Of Uranium.. 24

Nursan Akıncı
Urban City Life And Its Impact On Climate Change..................... 26

Turan Bayrm
The Future Of Renewable Energies In Gulf Coutries.................... 28

Orhun Ege Cansaran
Turkey Has No Need To Build Energy Plants Instantly.
Forcasts Mostly Wrong!... 30

Özgün Evrim Sayılkan
The Energy Potential Map Of Turkey...................................... 32

Rustam Guliyev
What If Neanderthals Come Back From Extinction?..................... 34

Seyedehshahrzad Seyfafjehi
Energy Resources On Mars.. 36

Büşra Yücel
Shop With A Grain Of Salt 38

Ezgi Altınöz
Climate Change Is On Our Plate... 40

Naz Alara Erbek
Carbon Capture Storage: Decrease Emissions –
Increase Efficiency... 42

Sardar Talal Khalid
A Letter To Oblivious Humans... 45

Feyza Yılmaz
No More Gasoline Cars? The World Is Turning To
Alternative Fuels - Turkey Is Not Falling Behind........................ 46

Onurcan Boran
How The World And Turkey Sees Electrical Vehicles................... 48

About the Authors... 51

THE IMPACTS OF GRETA THUNBERG ON ENVIRONMENTAL COMMUNICATION IN GERMANY

Lutz Peschke*

When Greta Thunberg started her protest outside the Swedish parliament in August 2018 and fought for immediate cautions to combat climate change, she was fifteen years old. Her school strike every Friday attracted media attention. Her media publicity resulted in a big school strike for climate movement. Since March 2019, all around the world in more than 100 countries, around 1.4 million students protest every Friday for the future of our habitat. The #FridayForFuture movement was born. Greta Thunberg exploited her popularity to give a voice to environmental activists. She gave a speech within the scope of the United Nations Climate Change Conference (COP24) in December 2018 in Katowice/Poland and other important meetings. But the feedback in politics, academia, media and society was ambivalent and should be described shortly in the case of Germany.

On the one hand, German politicians appreciated the engagement of young people for the environment. Chancellor Angela Merkel stated that the politics can reach the goal only, if they are supported by the society and that she supports the #FridayForFurture activities accordingly (Welt 2019, t=15s). The Federal Minister of Justice Katarina Barley's statement is similar and she commented that the voice of the youth should be strengthened. Therefore, she is in favour of voting rights from minimum age of 16 (ibid., t=41s) considering that 15.5 % of the electorates are between 18 and 30 but 20.7% are older than 70 (Der

---

* Asst. Prof. Dr. Dr. Lutz Peschke, Department of Communication and Design, Bilkent University, Ankara/Turkey

Bundeswahlleiter 2017). It reveals that the retired citizens in Germany have more influence on shaping the future then the youth. On the other side, the politics reacts refusing and distrustful on the youth activities and protests. Lindner a politician of the liberal party (FDP) posted a tweet that he likes and welcomes the political activities of the students but nobody can expect that young people understand the global context of the technical meaningful and economical feasible. This is the business of professionals (Lindner, 2019). He reduces the young people to the role of the participating audience while the actors in the arena are players from politics, industry and academicia. With this comment, he neglects the existing model of Quadruple Helix collaboration where the 'media-based and culture-based public' and 'civil society' is the fourth player in the collaborative process. According to Carayannis et al. the "natural environments of society and the economy also should be seen as [fifth] drivers for knowledge production and innovation" inside of a quintuple helix (Carayannis et al. 2012). Kramp-Karrenbauer, chair-lady of the conservative (CDU) complained that she would find it more convincing, if the students would not only fight for the climate during the class time but in their leisure time (phonix 2019). TV science moderator and medical doctor Hirschhausen commented Kramp-Karrenbauer's statement that he cannot remember that pilots, locomotive drivers or unionists proceeded their strikes in their leisure time. The idea to request that our society needs a change needs the pressure of #FridayForFuture strikes (Jung 2019, t=2037s). Additionally, several comments of politicians present the 'FridayForFuture' activities in the light of conspirations. Merkel stated within the scope of the Munich Security Conference in February 2019 in the context of hydrid warfare of Russia in the internet, that it is difficult that all German kids are able to get the idea for a protest at the same time without any influence from outside. Campaigns can be organized much easier with help of the internet (AG Kinder- & Jugendrechte 2019).

Since many years, scientists agree that the industrial revolution in the 19th century increased the impact of humankind on the climate change exponentially. Especially because of the high anthropogenic emission of carbon dioxide, global climate change is not anymore predominantly caused by natural behaviours like many millennia before. Therefore, Crutzen assigned the term "Anthropocene" as a new geological epoch to communication the human-dominated era which supplemented the Holocene, the warm period of the past 10-12-millennia (Crutzen 2002). The big transformation from productions in guilds and small manufactories which were commonly family business to industrial processed led to an energy need which was covered by fossil fuels. The problematic impact arose from the fact that carbon stored from millions of years of photosynthesis were emitted by combustions. While carbon dioxide which is emitted during the generation of energy captured from plants can be recycled by photosynthesis of renewable resources, coal, gas and oil emits additional carbon dioxide without any potential of compensation (Steffen et al. 2007, 616). After the critical and sometimes polemic comments and actions against the #FridayForFuture activities, the academic world reacted prompt. Hagedorn et al. founded the initative #Scientists4Future and wrote a petition where they confirmed that the "concerns [of the #FridayForFuture activists are justified and supported by the best available science. The current measures for protecting the climate and biosphere are deeply inadequate" (Hagedorn et al. 2019). On the one hand, the students got the protection of over 26.000 leading German scientist against abuses and fake news, on the other hand the scientists set the focus on the important discourse of how to initiate urgently needed changes and activities in favour of our habitat on Earth, away from useless discussions and polemic comments. Some scientists and science moderators organized a Federal Press Conference in Berlin to announce the implicit support of the #FridayForFuture activities. Maja Göpel, political economist, expert of climate politics and the general secretary of German Advisory Council on Global Change stated that the society discusses a structural change within the scope of digitalisation which is directly accepted. It is forced by the trade market, even although 25% of the jobs will disappear. It is accepted and celebrated as an agenda of progress. But why don't we have the same power to renew our climate technologies, way of agriculture, and way of mobility. Why do we celebrate digitalisation modernisation and progress including all collateral damages which is of course balanced by the politics while technological developments to prevent climate change is repelled until the last minute, although there will be damages in much higher orders. Göpel concludes that this question can be answered only according to political economical interests, hegemonies and constellation of power (Jung 2019).

The activities initiated by Greta Thunberg generates media coverages and reactions which reveals a big abyss between traditional and digital media and their recipients. The strongest impact had the influencer Rezo. He runs the YouTube channel "Rezo ja lol ey" where he normally talks about music and everyday life practices, plays games with guests, etc. But with his critical video about Article 13 of the Directive on Copyright in the Digital Single Market (Rezo ja lol ey 2019a), he attracted critical attention among in the society beyond the digital natives. But he became famous with a high amount of press coverage in traditional media though his video "Die Zerstörung der CDU" (engl.: the destruction of CDU) (Rezo ja lol ey 2019b). He criticizes 51 minutes, topics about winners and losers of the government's politics, the climate crisis, the ethics of war and the relation to the USA and other topics. The explosive power of the video was that he published the video some weeks before the election of the European Parliament 2019. He finished the video with the plea not to vote for AfD, CDU, SPD and FDP without giving recommendation to other parties but with the plea for going to the poll. Especially his part about the climate crisis was based on excellent research and was qualified by correct and reliable references and journalistic sources. Politicians, especially from the conservative party CDU reacted in a way which revealed the lack of understanding of influencer activities. Kramp-Karrenbauer blamed Rezo's spin and propaganda purposes and ask for regulation of digital media during election campaigns. The digital community reacted with shitstorms after her comments (Tagesschau 2019). Kramp-Karrenbauer's and other feedbacks revealed the digital abyss between generations. Unfortunately, the feedbacks dealt mainly with the way of Rezo's communication and not with the topics itself. The feedback of politics and media reveals that the way of peer communication of young people is obviously not understood. But an effective quintuple helix collaboration needs the integration of young activists in all areas.

The described problem is not a German problem but it shows a global lack on understanding on the one hand and a need of peer-learning activities on the other hand. In this context the project "POWER: Empowerment of Youth on Renewable Energy for Sustainable Societies" which is funded by the Erasmus+ Program of the European Union (2017-3-TR01-KA205-048402) aims to encourage young people for a critical dialogue about renewable energy as a contribution to combat climate change. The core issue of this project is a training document for young trainers which will be developed and presented on the 4th Transnational Project Meeting at Vienna University of Technology in August 2019. The central aspect is the peer-learning methodology (PLM). The main aspect of PLM is that students learn from others (peers) by actively engaging in large trainee groups to experience practical skills, in combination with appropriate and real time feedback in order to improve their deep understanding and learning of real-world aspects. This book presents contributions of students of the course "Science Writing and Journalism" of the Department of Communication and Design at Bilkent University in Ankara. They created posters and popular texts about renewable energies and related topics. The contributions are part of an exhibition which will be presented on Renewable Energy Events in five Turkish cities between October 2019 and January 2020. During these events students of Bilkent University will apply the peer learning concept.

Finally, I would like to thank our coordinator Ankara Provincial Directorate of Environment and Urbanization, the international partners Open Evidence in Barcelona/Spain, Vienna University of Technology in Austria, iserundschmidt – agency for science communication in Bonn/Germany, as well as our Turkish partner TERA Ankara and my colleagues and students from the Department of Communication and Design at Bilkent University in Ankara for the fruitful and inspiring collaboration. Last but not least, I would like to thank the Turkish National Agency for the big support.

Ankara, July 2019
References

AG Kinder- & Jugendrechte . 2019. "Angele Merkel betrachtet 'Fridays for Future' als hybride Kriegsführung". Accessed July 26, 2019. youtu.be/Us4jnNmW3xU

Carayannis, Elias G., Thorsten D Barth and David F.J. Campbell. 2012. "The Quintuple Helix innovation model: global warming as a challenge and driver for innovation". *Journal of Innovation and Entrepreneurship*, 1: 2.

Cohen, Ilana and Jacob Heberle. 2019. "Youth Demand Climate Action in Global School Strike. Harvard Political Review". Accessed July 26, 2019. http://harvardpolitics.com/united-states/youth-demand-climate-action-in-global-school-strike/

Crutzen, Paul J. 2002. "Geology of Mankind". *Nature* 415(3): 23.

Der Bundeswahlleiter. 2017. Bundestagswahl 2017: 61,5 Millionen Wahlberechtigte. Accessed July 26, 2019. https://www.bundeswahlleiter.de/info/presse/mitteilungen/bundestagswahl-2017/01_17_wahlberechtigte.html

Hagedorn, Gregor, Peter Kalmus, Michael Mann, Sara Vicca, Joke Van den Berge, Jean-Pascal van Ypersele, Dominique Bourg, Jan Rotmans, Roope Kaaronen, Stefan Rahmstorf, Helga Kromp-Kolb, Gottfried Kirchengast, Reto Knutti, Sonia I. Seneviratne, Philippe Thalmann, Raven Cretney, Alison Green, Kevin Anderson, Martin Hedberg, Douglas Nilsson, Amita Kuttner, and Katharine Hayhoe. 2019. "Concerns of young protesters are justified". *Science* 364 (6436): 139-140.

Jung, Tilo. 2019. BPK: "'Scientists for Future' zu den Protesten für mehr Klimaschutz - 12. März 2019". Accessed July 26, 2019. youtu.be/OAoPkVfeTo0?t=2037

Lindner, Christian (@c_lindner). 2019. Twitter, March10, 2019. https://twitter.com/c_lindner/status/1104683096107114497?ref_src=twsrc%5Etfw%7Ctwcamp%5Etweetembed%7Ctwterm%5E110468309 6107114497&ref_url=https%3A%2F%2Fwww.fr.de%2Fpolitik%2Ffridays-for-future-christian-lindner-kritisiert-schuelerdemonstrationen-gegen-klimawandel-11842275.html

Phoenix. 2019. "JU-Deutschlandtag - Rede von CDU-Chefin Annegret Kramp-Karrenbauer am 16.03.19". Accessed July 26, 2019. https://youtu.be/9GMiDy0LZQ4?t=585

Rezo ja lol ey. 2019a. "Ich entlarve Propaganda zu Artikel 13". Accessed July 26, 2019. youtu.be/iNpB73CAdL8.

Rezo ja lol ey. 2019b. "Die Zerstörung der CDU". Accessed July 26, 2019. youtu.be/4Y1lZQsyuSQ.

Steffen, Will, Paul J. Crutzen and John R.McNeill. 2007. "The Anthropocene: Are Humans Now Overwhelming the Great Forces of Nature?" *Ambio* 36(8): 614–620.

Tagesschau. 2019. "tagesthemen 22:15 Uhr, 28.05.2019". Accessed July 26, 2019. youtu.be/VvSQvfL-DHo?t=35.

Welt. 2019. "FRIDAYS FOR FUTURE: Kanzlerin Merkel lobt Schülerdemos für den Klimaschutz". Accessed July 26, 2019. youtu.be/KCriQVHQ7b4.

ENVIRONMENTAL LAW CLINICS: TOWARDS A NEW ERA IN LEGAL EDUCATION

Seldağ Güneş Peschke[*]

In the last years, the law school students are encouraged to have practices on legal subjects, before they are graduated from the university. In this sence, law clinics are founded to give the students legal expertise and education under the supervision of academicians and practisioners from public and private sectors.

One of the aims of the law clinics is to prepare students to practice law. These law clinics provide training directly relevant to many law students' future practices. These law clinics guide law students in a wide approach that they are well trained in legal science besides they become ethical lawyers. In this sense, the law clinics expand the access to legal system in highly regulated areas (Babrich 2013, 45-47). In the law clinics the students are motivated to discuss legal issues with the academicians and the practitioners.

Clinical experience supports students to become skilled professionals who are able to develop arguments, explore strategies and communicate effectively with clients, courts, scientific experts and opposing parties. In the law clinics, students receive hands on training in the practice of law. They sometimes represent real clients under the supervision of their professors. Most of the clinics provide free legal services to clients with low budget (Ashar 2007, 356).

Under these circumstances, for the development of environmental law, there is a big need to train the next generations. The environmental problems are increasing each day all around the World. The countries are

---

[*] Prof. Dr. Seldağ Güneş Peschke, Ankara Yıldırım Beyazıit University, Faculty of Law.

trying to regulate environmental issues in the national level considering the international principles. Effective environmental law and legal systems throughout the World is important not only for the economic development but also for the environmental sustainability of the future. There is a need of cross border collaboration and transnational networks for the development of environmental law. So, all national and international regulations can be set up together under a global environmental law. (Yang et al. 2009, 617) Global environmental law should be considered as a branch of law which is effected from both public and private law principles in the framework of comparative law. Transnational legal processes, legal traditions, international networks, judicial cooperation across borders should also be considered in the formation of global environmental law.

In the last years there is an increase in the number of environmental law clinics all around the World. According to Schroek, University of Oregon and the University of Colorado in collaboration with the National Wildlife Federation were the first universities which formed environmental clinics during 1975 and 1978. Today, approximately 20 percent of the law schools in the USA operate environmental law clinics which serve an important role in the protection of natural resources and in the prevention of pollution (Schroek 2016, 2).

These law clinics guide the law students to become ethical lawyers, so these students have the access to legal system and they afford legal help on environmental issues. These law clinics are effective in the policy making process as well. For example, University of Michigan School of Law students who took part in the environmental clinics assisted in drafting the Michigan Environmental Protection Act of 1970 (MEPA) language. The work of these students effects positive changes to federal and state environmental laws. The University of Maryland Law School Environmental Law Clinic (Maryland Clinic) has also been engaged in environmental justice work. In Georgetown University environmental justice clinic has been formed just to address and remediate instances of environmental inequity (Bobcock 1995, 44; Schroek 2016, 9).

China takes as an example the environmental clinics in USA. For instance, University of Zhongshan, conducts surveys of residents, enterprises and government agencies for policy making. The students from Remnin University environmental clinic educate the public about environmental laws. (Schroek 2016, 13)

With the Paris Agreement, for time all nations are brought into a common cause to undertake ambitious efforts to combat climate change. Paris Agreement's main aim is to strengthen the global response to the threat of climate change. But to undertake these ambitious efforts, there is a need to inform the next generation about legal framework of Environmental Law in their countries and to encourage them to create concepts for a sustainable workaday life.

European environment policy has evolved significantly since the 1970s. The protection of environment is one of the most supported policy areas by EU citizens, who recognise that environmental problems go beyond national and regional borders and can only be resolved through concerted action at EU and international level. Under these conditions, EU countries are perfect examples to make young people sensitive to environmental issues, waste management to develop policies that attract youth.

Law schools start to give importance to the environmental law clinics in the last ten years, to lead the students to experiential learning from theories to practices. In this framework, the Environmental Law Clinics establish the interaction between law students and the practitioners by increasing the skills of law students in Environmental Law and encouraging them about concepts how to make the way of life more sustainable according to avoiding waste, replacing materials to more ecologic ones, protecting nature in rural and urban areas, etc. Since the Paris Agreement needs the global interaction of all countries, Environmental Law Clinics are supported by legal experts, environmental activists, academicians, non-governmental organisations, and the States.

In the clinics, students examine specific legal cases and questions dealing with international environmental protection and regulations in a comparative way. The students work together with the academicians and representatives of the organizations. Clinical education offers students, experience in the practice of environmental law.

In weekly seminars, students find time to discuss strategic decisions of the environmental issues and consequences of some environmental cases. During the lectures, the law students have the chance to work with local, regional, and national environmental and community organizations. Besides, they try to find solutions to critical environmental challenges in the big cities, like waste management; the process of prevention, control, monitoring, collection, transportation, reuse, recycling, disposition etc.

In each university the concept and curriculum of environmental law clinics can be different, but the main idea of them remain generally the same: the protection of environment. The lectures in the clinics mainly focus on the international and national regulations on environment, climate change and waste management from the perspective of fundamental principles of environmental law.

The Environmental Law Clinics create a highly innovative method of teaching as it includes the education of law students how to become ethical lawyers and represent clients. Besides it gives a basic legal and scientific knowledge on environment. During the lectures, the students have the chance to attend the seminars, conferences, workshops within the curriculum of the lecture.

As a result, environmental law clinics encourage law students in their future profession, in the field of environmental law. When the graduates become policy and decision-makers in governments or in business life, or work as consultants, researchers and academicians, or as lawyers, they take care of environment and sustainable development accordingly. By learning the legal situations, within the scope of Environmental Law, they have the chance to create concepts and projects which contribute to the Paris Agreement within the United Nations Framework Convention on Climate Change to strengthen the global response to the threat of climate change.

Bibliography

Babcock, Hope. 1995. "Environmental Justice Clinics: Visible Models of Justice" *Stanford Environmental Law Journal* 14(1): 3-60.

Babich, Adam. 2013. "Twenty Questions (and Answers) about Environmental Law School Clinics". *Professional Lawyer* 22(1): 45-54.

Dodge, Amanda, and Gemma Smyth. 2018. "Learning, Teaching & Practising Systemic Advocacy in Legal Clinics: A Conversation". *Journal of Law and Social Policy* 29(47): 47-66

Lin, Yanmei. 2016. "Environmental and Biodiversity Law Clinic at Southwest Forestry University: A New Environmental Law Clinic Model in China". *Vermont Journal of Environmental Law Vermont Law School* 18(1): 18-35.

Shanahan, Colleen F., Jeffrey Selbin, Alyx Mark, and Anna E. Carpenter. 2018. "Measuring Law School Clinics". *Tulane Law Review* 92(3): 547-586

Schroeck, Nicholas J. 2016. "A Changing Environment in China: The Ripe Opportunity for Environmental Law Clinics to Increase Public Participation and to Shape Law and Policy". *Vermont Journal of Environmental Law Vermont Law School* 18(1): 1-18.

Yang, Tseming, and Robert V. Percival. 2009. "The Emergence of Global Environmental Law". *Ecology Law Quarterly* 36(3): 615-664

GEOTHERMIE IS SUPER-HOT AND AS BEAUTIFUL AS EVER

M. Mert Örsler[*]

Geothermal energy is just like a beautiful girl with her charming red dress on a cocktail party, waiting for his cavalier: Turkey. But Turkey seems to have a date with fossil fuels nowadays.

The country is ranked 3rd in the world regarding energy dependency; Turkey's interest in energy for the industrial and everyday usage is not a secret. But does it have to be met by importing Iranian gas? Can national and renewable resources meet Turkey's needs? To a certain extent, the answer is yes; the country can meet the needs of energy thanks to geothermal power! For the Royal Danish Consulate General Anette Snedgaard Galskjøt, the country can easily cut the imports of gas by 20 % using geothermal power, which can cater for 30 % of the country's heating needs. Turkey has already taken some bigger steps towards a geothermal future in İzmir, Denizli, and Çanakkale. But this should not sound sufficient; the country is ranked 7th, right after Iceland, across the globe regarding the geothermal potential. In Denmark, a country that is not even listed among the top geothermal countries unlike Turkey, almost two third of the population uses renewable energies including geothermal energy for heating. This is even closer to 100 % in Copenhagen. Despite the great potential of geothermal energy, the country insists on importing oil and gas which reached the biggest percentages of the annual energy import: 44% and 35 %. On the other hand, geothermal power is interestingly the least used energy source for electricity and heating in the country: 0.5 %. This situation can change if Turkey starts realizing the geothermie, super-hot and as beautiful as ever, waiting for far bigger steps from Turkey which might, at least, hinder the country's long-term relationship with the fossil fuels.

---

[*] M. Mert Örsler, Department of Communication and Design, Bilkent University, Ankara/Turkey

GEOTHERMIE IS SUPER-HOT AND AS BEAUTIFUL AS EVER!

Turkey mostly meets her needs by importing oil and gas. The use of geothermal energy can easily reduce this dependency on non-renewables for, especially, electricity and heating.

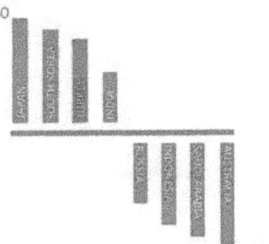

FREE TURKEY FROM THE FOREIGN FOSSIL FUELS
Energy dependency means to what extent a country's economy is dependent on the imports so as to satisfy its needs. The graph shows the top 4 and bottom 4 countries in terms of EG.
■ Source: Eurostat

ITS RUSSIA'S OIL AND IRAN'S GAS
The pie chart shows the percentages of Turkey's imports annually. Oil and gas is imported most.
■ Source: Eurostat

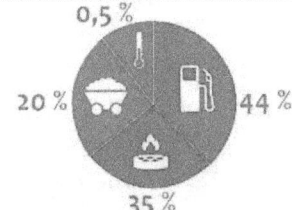

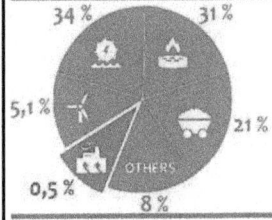

DON'T BE FUELISH
The pie chart demonstrate the percentages of energy production for electricity and heating. Geothermal energy is the least one used although Turkey is ranked 8th with 635 MWe installed capacity for Selectricty and heating in the world.
■ Source: Eurostat

WITH LOVE FROM MOTHER EARTH
The table on the right hand side describes the major areas of usage regarding geothermal energy and some of the distinct heat in cantigrade degree.
■ Source: Eurostat

Heat	Area of Usage
20 °C	fish farms
40 °C	soil heating
60 °C	heating for greenhouse and livestock industry
80 °C	heating for urban and rural areas
100 °C	blighting organic metarials such as vegetables and canning
120 °C	heating for fresh/potable water
140 °C	sugar and salt industry
160 °C	lumber industry
180 °C	power generation-electricity

Credits: M. Mert Orsler

SCHEME GEOTHERMAL POWER

Design: Melina Diener, Marleen C. Schwalm
Concept: Manfred Schmidt, Ruben Düchting
iserundschmidt GmbH – Agency for Science Communication, Bonn/Germany

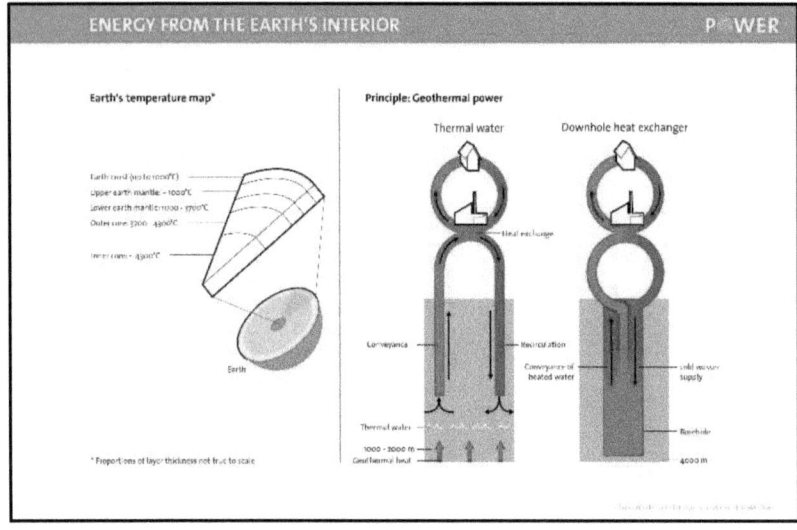

SCHEME WINDPOWER

Design: Melina Diener, Marleen C. Schwalm
Concept: Manfred Schmidt, Ruben Düchting
iserundschmidt GmbH – Agency for Science Communication, Bonn/Germany

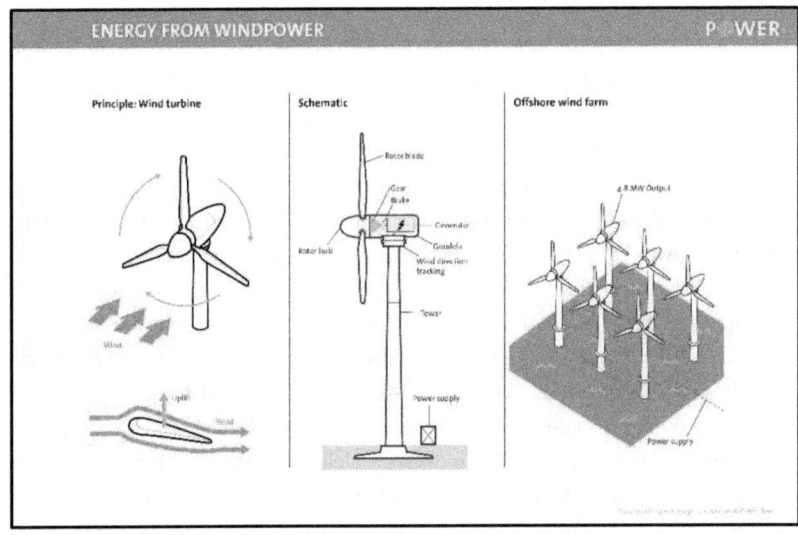

PROS AND CONS OF HYDROPOWER
Simge Sadak*

Among renewable energies hydropower is at the top after wind energy in terms of being the greenest. In fact, hydropower is the largest renewable energy source in the world getting ahead of the wind energy. Despite the huge Green House Gas (GHG) emission rates, hydropower decreases the damage by having one of the least GHG emission rate compare to the other energy sources.

Hydropower has a huge capacity of energy production that shouldn't be look aside. According to International Hydropower Association, U.S and Brazil has the most installed hydropower capacity in the world following the China. Today, China has over 341,000 MV installed capacity, U.S fallows China with 103,000 and Brazil has 100,000. Turkey also has 27,000 MV installed capacity in 623 hydropower plants and produces %20 of it's electricity from hydropower.

Due to the source that hydropower uses, which is the water, it is clean and reliable. Also as a renewable energy it avoids pollution and Green House Gas emissions. However, there are some disadvantages that should be take into consideration. Construction of these hydroelectric power plants block the fish migration and cause extinction of some fish species. In fact, few Chinese fish species are already endangered because of these plants. Another important problem that should be paid attention is that these hydropower plants sometimes cause floods due to the overtopping and this leads to serious destruction of existing systems such as jobs, settlements and also high number of deaths.

*Simge Sadak, Department of Communication and Design, Bilkent University, Ankara/Turkey

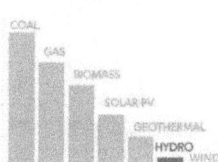

SCHEME HYDROPOWER

Design: Melina Diener, Marleen C. Schwalm
Concept: Manfred Schmidt, Ruben Düchting
iserundschmidt GmbH – Agency for Science Communication, Bonn/Germany

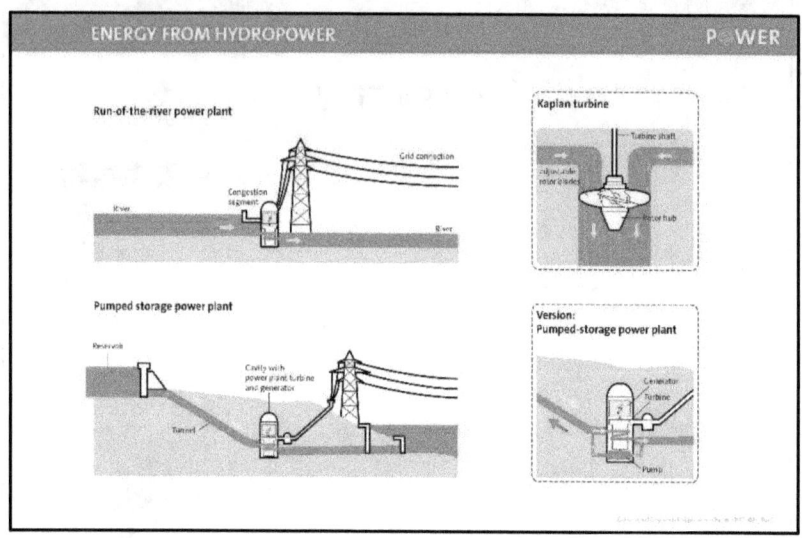

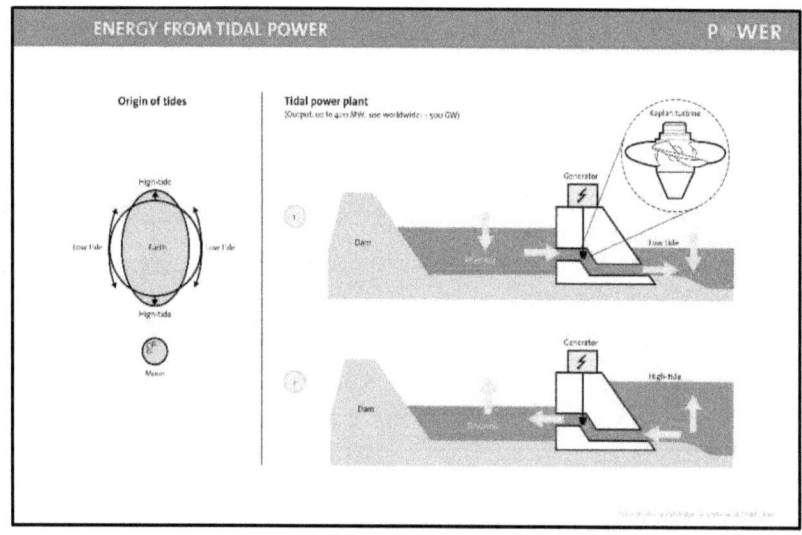

SCHEMES ENERGY FROM SUNLIGHT

Design: Melina Diener, Marleen C. Schwalm
Concept: Manfred Schmidt, Ruben Düchting
iserundschmidt GmbH – Agency for Science Communication, Bonn/Germany

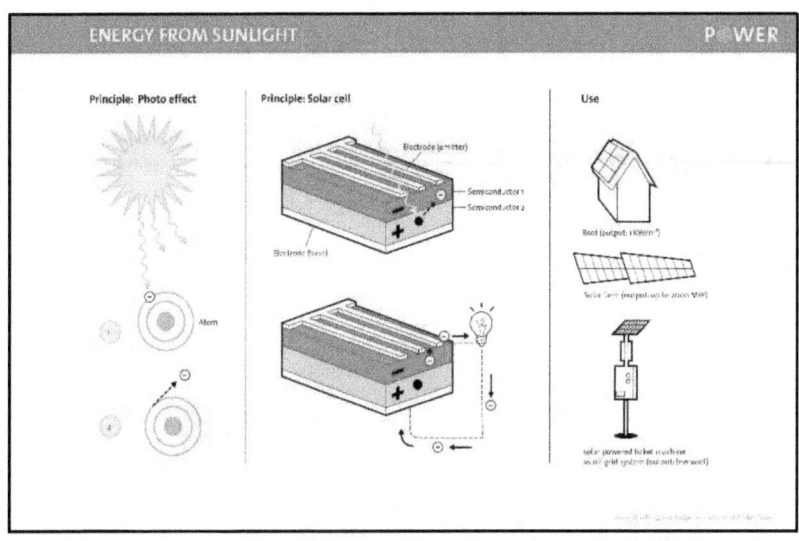

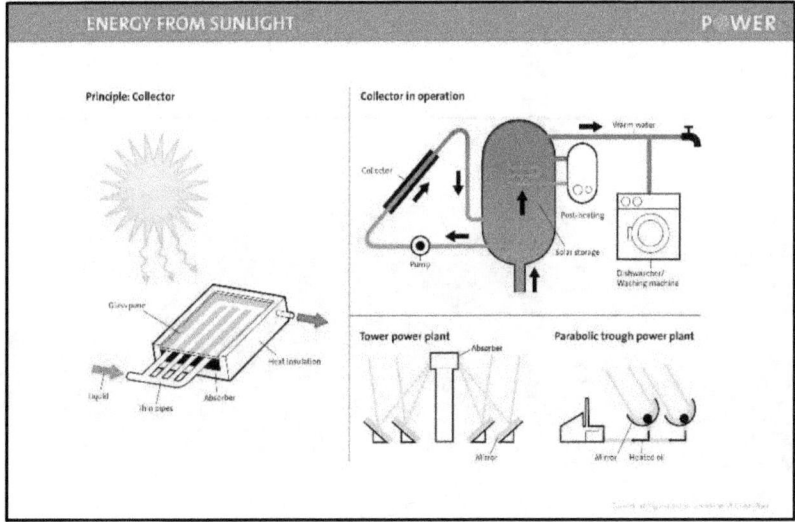

SCHEME BIOMASS

Design: Melina Diener, Marleen C. Schwalm
Concept: Manfred Schmidt, Ruben Düchting
iserundschmidt GmbH – Agency for Science Communication, Bonn/Germany

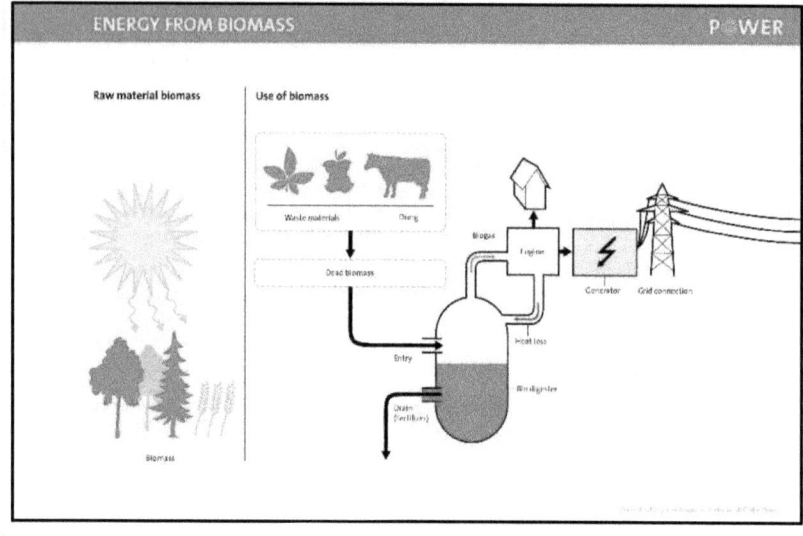

NUCLEAR POWER IS NOT WANTED IN TURKEY

M. Mert Örsler*

According to recent research, almost two-thirds of Turkish people are opposed to nuclear energy in the country. Nearly 90 percent of the participants of research living in Akkuyu, Sinop, and İğneada are worried about the potential of a Chernobyl-like nuclear accident.

Based on Dr. Pınar Ertör-Akyazı and her colleagues' research from Boğaziçi University, 2422 residents from urban Turkey were surveyed face-to-face. Their findings indicate that Turkish citizens endorse renewable energies while their opposition to the nuclear energy in Turkey is strong. Although only 7.2 percent shows their endorsement regarding nuclear power, 62.5 percent of people state that they are strongly opposed to nuclear power in Turkey. On the other hand, renewable energies such as wind energy and solar energy are endorsed by the same participants of the research; 70.2 percent ranked renewables as their first or second choice while only 4 percent as opposed to renewables.

According to the research of Emrah Akyüz from University of Leeds Sustainability Research Institute, people live in the reigns where the nuclear energy plants are constructing and will be built soon are explicitly opposed to the construction of a nuclear plant in their region. The research published in 2017, based on semi-structured interviews with 90 people of these locations, the local people are worried about the likelihood of an accident because, for instance, Akkuyu is an earthquake region. People living in Sinop, however, are concerned about the issues related to the problem of nuclear waste. Sinop's people are proud of the natural beauty of their region and worried about the unwanted potential of nuclear waste, which can be a serious threat to the touristic destinations of the city and the seacoasts.

The participant of both kinds of research is all worried about a nuclear future that Turkey has signed. For a country with a sorrowful past due to the Chernobyl accident whose devastating impacts on people living in Black Sea reign had been covered by local authorities for decades poses a big question regarding nuclear power plants in Turkey.

---

*M. Mert Örsler, Department of Communication and Design, Bilkent University, Ankara/Turkey

THE LONG LIFE OF URANIUM
Öykü Öncül*

After the discovery of Uranium, it become the main fuel for both nuclear reactors and raw material for nuclear weapons. Due to its characteristics of split into two different lighter fragments and its releasing energy process. Even though uranium consists of three isotypes: uranium-234, uranium-235, and uranium-238. All of them are radioactive. One of the most common isotope uranium-238 has a half-life about 4.5 billion years which means that the half the atoms will be decaying in that amount of time which are more beyond human time. Due to uranium-235 characteristics since it can maintain a chain reaction in which each fission able to produce neutrons to trigger other and fission process can sustained by itself without any external source of neutrons that's because it highly used in nuclear weapons and power plants that has 704 million half-life. On the other hand, uranium-238 even though could not sustain chain reaction by emitting alpha particles that are less potent by comparing other forms of radiation and the effects of the its gamma rays remains outside the body, uranium by considering the gamma-rays poses little health risks. However, if it is inhaled its radioactivity cause increased risk of lung, bone cancer and can cause damage to internal organs. Animal studies show that uranium has influence on both the developing fetus and reproduction which is the sign of its danger. Moreover, effects of the milling and mining uranium operations since it poses the increased lung cancer. Many of the Native Americans who worked in uranium mines, died because of lung cancer. Effects on water and land contamination still continue due to its both radioactivity and its quality of remain hazardous even after thousands of years.

---

*Öykü Öncül, Department of Communication and Design, Bilkent University, Ankara/Turkey

URBAN CITY LIFE AND ITS IMPACT ON CLIMATE CHANGE

Nursan Akıncı*

High buildings and their reflective glasses, asphalt on the roads, excessive population, and going to everywhere with our car have big impacts on climate change. NASA explains "extraordinary raining" in megacities as a result of the urbanization. The skyscrapers increase the temperature with their mirror glasses. Additionally, the urban soil is covered and sealed with the asphalt that the rainwater cannot be absorbed. The rainwater rises to the atmosphere again. The result is the heavier and abnormal raining. Since İstanbul is the 4th biggest city in the world and has approximately 200 skyscrapers and more than 5 thousand high building, these abnormal raining has seen in recent years as well.

Furthermore, the population is responsible for extraordinary raining in terms of increasing effect on climate change as well. In İstanbul live more than 14 million people, use electricity, and cars in a highly excessive way. Human's developed lifestyle are a crucial reason for the climate change which may not be seen as unexpected.

---

*Nursan Akıncı, Department of Communication and Design, Bilkent University, Ankara/Turkey

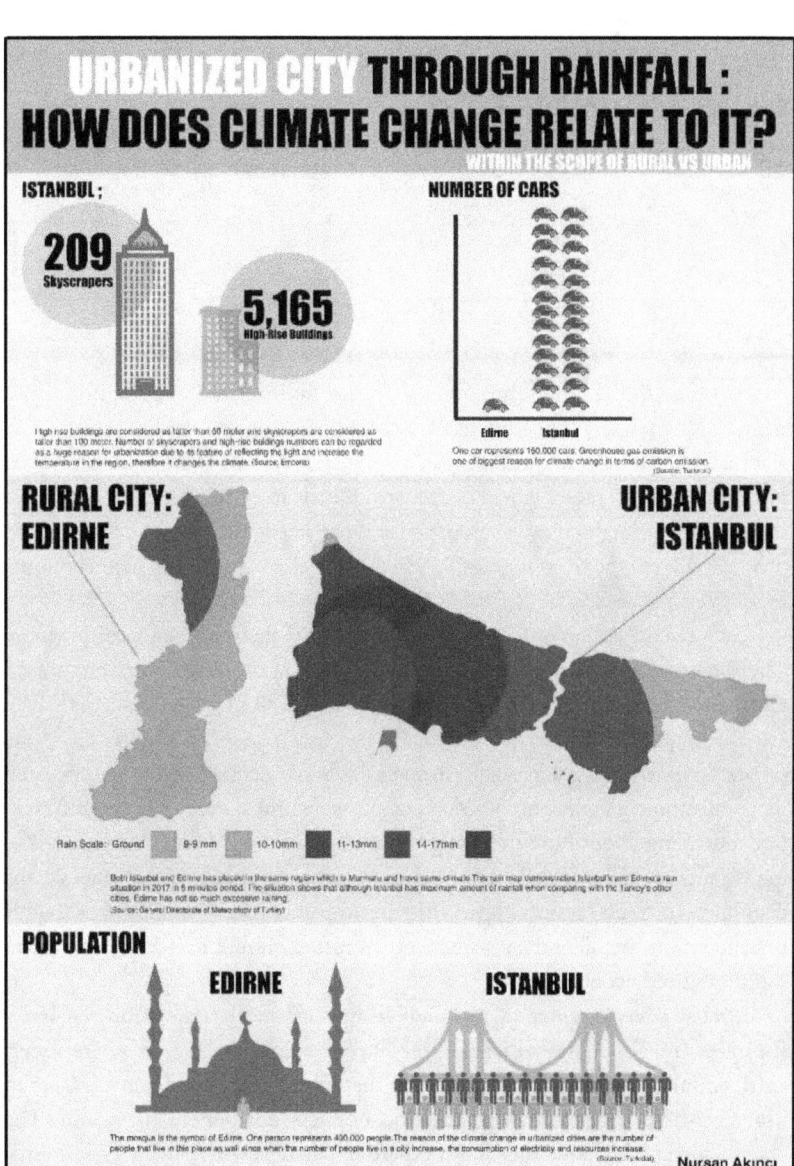

THE FUTURE OF RENEWABLE ENERGIES IN GULF COUNTRIES
Turan Bayram*

Not only in the long term, but also in the short term, Gulf countries need renewable energies, since their electricity consumption is increasing. The electricity consumption in the region have increased 12.4% from 2005 to 2009 in average which is the highest in the world. Instead of producing energy from hydrocarbons for their own usage, it is most likely that they will choose to produce this energy from renewable which will help them sustain their economy in long term as well.

Gulf countries has a strong economy due to their geographical position: They have the most fertile lands to produce oil and gas. To open up the subject, Gulf countries hold 29% of all oil reserves, and 22% of all gas reserves which are the highest ratio in the world. These fertile resources gave them the opportunity to raise golden cities in the middle of desert. They indeed enjoyed this opportunity for the past 50-60 years, yet the world started to have a different understanding about how the energy should be produced and consumed. The importance of renewable energies is increasing; so it will affect directly the economy of Gulf countries in the future. Since their economy is highly dependent on the oil and gas sails; Gulf countries should find a way to sustain their stabilized economy.

Almost every country in the Gulf has a long-term strategy on the issue. Emirates and Saudi Arabia leading the change across the region, yet relatively small countries have plans as well. As a matter of fact, Gulf countries have to use renewable energies due to their economical position in the world. The change seems a bit slow since it is too hard to change everything in a short time, yet the plans seems as promising as it can be.

---

*Turan Bayram, Department of Communication and Design, Bilkent University, Ankara/Turkey

THE FUTURE OF RENEWABLE ENERGIES IN GULF COUNTRIES

FACTS AND FIGURES

UNITED ARAB EMIRATES
- 7% of installed capacity in Abu Dhabi till 2020
- 25% of electricity supply in Dubai till 2030
- 75% of electricity supply in Dubai till 2050

1 Gulf countries hold 29% of all oil reserves. They also have 22% of all gas reserves across the globe. This rate is the highest in the world.

SAUDI ARABIA
- 9.6 GW mix of wind, solar and waste to energy till 2023
- reduce electricity consumption by 8% till 2021

BAHRAIN — 5% of installed capacity until 2020

2 The electricity consumption in the Gulf countries had increased rapidly; 12.4% from 2005 to 2009. This rate is much larger compared to the world.

QATAR
- 1.8 GW solar power till 2030
- 20% reduction in per capita electricity consumption

3 It's more likely for them to use renewables, since hydro-carbons is the most important thing they export.

KUWAIT
- 5.7 GW Concentrated Solar Power
- 4.6 GW Solar Power
- 0.7 GW Wind Power, all due 2030

OMAN — Currently preparing a long-term energy strategy

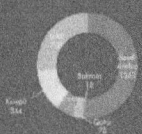
Cumulative installed solar capacity in Gulf, 2018, total: 2660 MW

In 2008, Masdar City, located near by Abu Dhabi, was built as the most sustainable eco-city in the world.

Masdar inaugurate Shams 1, one of the world's largest Concentrated Solar Power (CSP) plants.

Cumulative installed wind capacity in Gulf, 2018 total: 537 MW

The Carbon Capture, Usage & Storage (CCUS) project captures up to 800,000 tonnes of Carbon Dioxide (CO_2) from local region.

12896 kilowatt is consumed per capita in average throughout the Gulf, population is around 50 million (1 megawatt = 1000 kilowatt)

TURKEY HAS NO NEED TO BUILD ENERGY PLANTS INSTANTLY. FORCASTS MOSTLY WRONG!

Orhun Ege Cansaran[*]

Turkey's energy politics has been changed contrary in order to make easier to build new energy plants since the beginning of the new millennium. Turkey was getting in an available environment to provide cheap and generous credits. On the other hand, global economy struggled with a huge crisis in 2008 and its effects is perceived whole around the world. Turkey's growth speed slowed down from around 7% to 3% so it affected the forecasts about the country's future. Electric consumption growth has a correlation with economical growth. Consequently, Turkey isn't growing as like last decade but the energy production policies still based on old fore-casts. That's why the ratio between Peak Power Demand and Installed Capacity is 55,93%

Turkey has no need to build energy plants instantly, forecasts mostly wrong! Turkey can produce more energy than it needs but the plants should be off so it may call as public nuisance. The capacity is already higher than needs of Turkey at the rate of 1.7 times.

Does Turkey need that much energy plants? Forecasts say yes, results say no! Energy plants investment of Turkey has a consistently rise for a decade and approximately 5900 MV installed capacity codify the system every year. These investments cost over 6 billion dollars per year for over 1500 plants and most of the investment is met by foreign financing with foreign currency. Because of the exchange risk, the corporations must sell the outcome with a stable price in order to redress the balance. Otherwise, corporate ones will fail with unpaid credits. That's why, government authorities chose the private companies to buy energy despite the higher prices.

Official forecasts are mostly wrong to predict more than 2 years. Phases of construction and permissions takes years so forecast is cruxes to make right decisions. On the other hand, official forecast of Turkey doesn't accurate if forecast situates over two years so most of the new plants don't profitable businesses. Government tackles that issue with warranty provisions but it also occasions public loss. Be-cause, hydro plants of public corporations don't work. Despite all, many numbers of natural gas combined cycle plants are shut down or replace since August 2018.

---

[*]Orhun Ege Cansaran, Department of Communication and Design, Bilkent University, Ankara/Turkey

POWER GENERATION
OF TURKEY
An Analysis on Guesses and Facts

GDP OF TURKEY
851 BILLIONS $

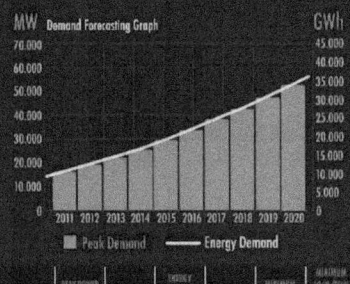

 MONEY MOSTLY COMES FROM FOREIGN BANKS WITH FOREIGN CURRENCY

102,5 BILLIONS $= INTEREST PAYMENT OF TURKEY (2018)

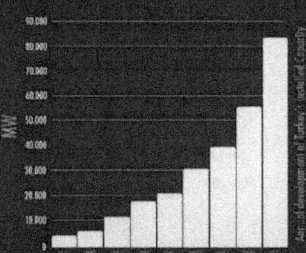

Turkey has no need to built new energy plants instantly, forecasts mostly wrong!

Only in 2017, 1583 energy plants was opened that have 6089 MV installed capacity with investing around 6,2 billion $

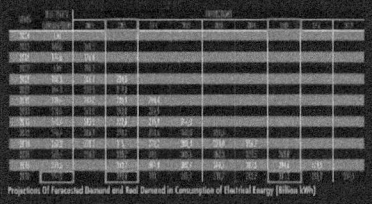

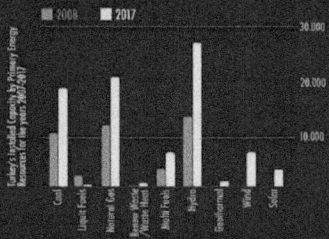

$$\%55{,}93 = \frac{\text{MAXIMUM PEAK DEMAND}}{\text{CAPACITY}}$$

References
TEIAS - Turkish Electricity Transmission Corporation
Ministry of Energy and Natural Resources

Orhun Ege CANSARAN

THE ENERGY POTENTIAL MAP OF TURKEY
Özgün Evrim Sayılkan[*]

Turkish energy policy is concentrated on 3Es, which compromise of energy, economy, and environment refers to the assurance of reliable, sufficient and timely manner energy production. However, the distribution of energy production to resources in the recent year causes suspicion about whether Turkey follows this policy successfully or not.

The distribution of energy production to resources in the recent year is as below: %35.05 natural gas, %19.56 import coal, %19.38 hydropower, %16.42 lignite, %6.31 wind, %2.16 geothermal, %0.75 biomass, and the others.

Turkey is a relatively lucky country thanks to the fact that almost all renewable energy resources including geothermal, biomass, solar, wind, and hydropower are available. Having an energy-rich geographic location brings Turkey more responsibilities about recognizing the potential of clean, domestic and economic resources and taking the advantage of them in an efficient way as it is indicated in 3Es policy. However, the current state of energy production brings a question whether Turkey uses all potential of renewable energy sources which are sufficient to energy production or not.

---

[*] Özgün Evrim Sayılkan, Department of Communication and Design, Bilkent University, Ankara/Turkey

THE RENEWABLE ENERGY MAP OF TURKEY

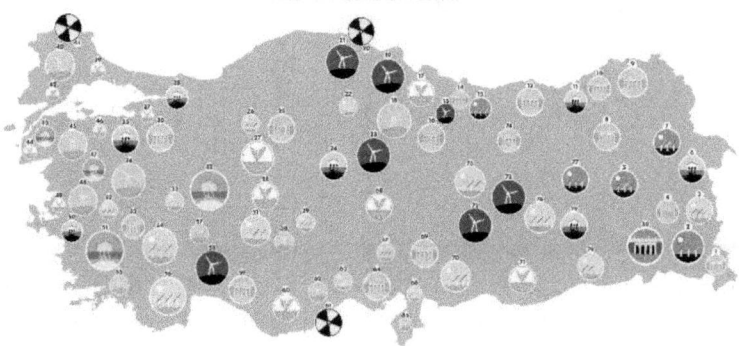

HOW TO PLAY ❓

It can be played with 2 or 3 people.

Roll the dice and move your game piece according to the number.

Check out the guidance what you have on your number.

Perform what the number says on the guidance.

Repeat until you arrive the last number.

The person who arrives the biggest number wins the game.

IMPORTANT NOTES ❗

Each icons refers to actual potential.

Icons that have currently active usage are colored, the others are in grayscale in order to show unused potential.

Numbers are ordered with the aim of traveling all around of Turkey.

THE PURPOSE ❗

To uncover the situation of renewable energy usage in Turkey by portraying used and unused renewable energy potential.

 Hydropower   Geothermal Energy   Biomass   Solar Energy

 Wind Energy   Nuclear Energy   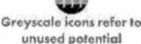 Greyscale icons refer to unused potential

 The potential increases or decreases according to size of the icons

http://www.enerjiatlasi.com   http://www.yegm.gov.tr   Özgün Evrim Sayılkan

WHAT IF NEANDERTHALS COME BACK FROM EXTINCTION?

Rustam Guliyev*

The Earth - a beautiful planet that popped up about 4 billion 400 million years ago, on its surface slowly the oceans formed and then the conditions for life appeared, well, but who will inherit this blue planet. Humans are probably one of the dumbest beings on Earth if you count them with a theory of survival, for thousands of years as Homo sapiens evolved; he does his best to destroy himself as a species. We began to change our planet so much since our inception that officially the world began to recognize the existence of the Anthropocene period due to the strong influence of man on the geological and biospherical structure of the Earth.

However, is this feature of the Homo sapiens a representative of only our species or is it just our genetic lineage starting with our ancestors Australopithecus? Do we really need Anthropocene to claim that it wraps our essence? So this is a story of a heroic war for our world, the war between monsters. Per contra, the behaviour of Neanderthals suggests otherwise. Their influence on the planet is still under our feet and their DNA is still in us.

Neanderthals were famous hunters during Pleistocene; even in winter within the large migrations of animals, they attacked them more frequently than ever; such as mammoths, Irish elks, and roes. Over time, the population of those animals declined. Funny enough, today, we cut a large population of huge mammals that they don't have time to reproduce. Neanderthals were so erudite in terms of fire that over time their work could grow into huge forest fires, which undoubtedly resembles us. Even the fact that we have from 1 to 4 % of the DNA of a Neanderthal is itself already enough to understand that we had the same interests in shaping the environment.

---

*Rustam Guliyev, Department of Communication and Design, Bilkent University, Ankara/Turkey

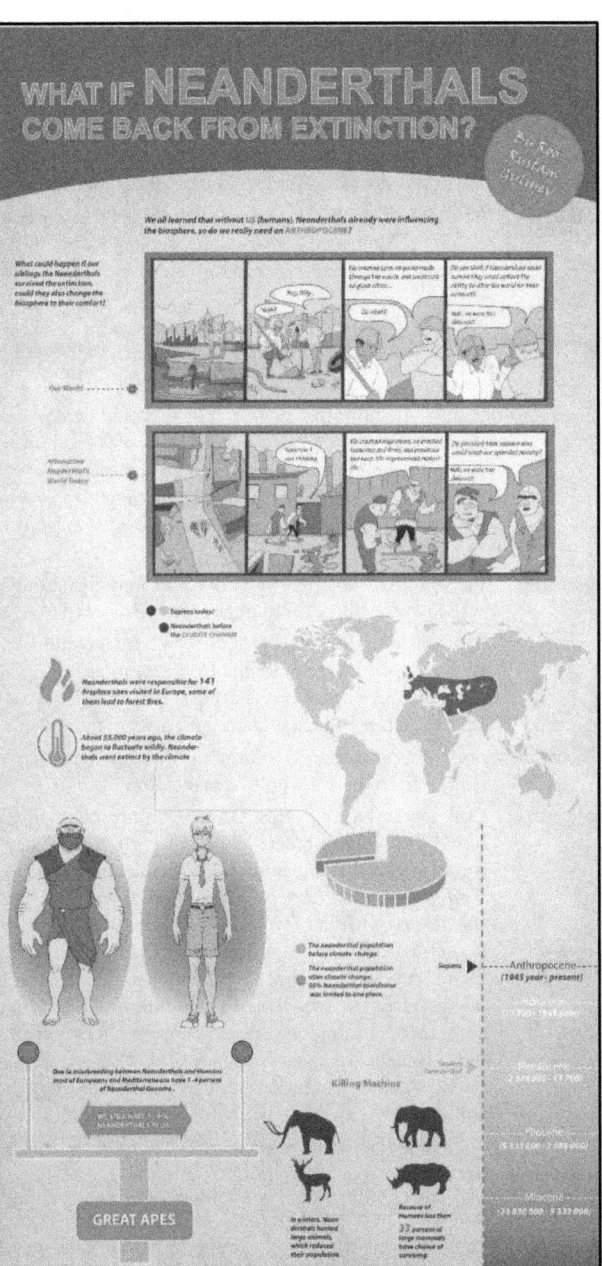

ENERGY RESOURCES ON MARS

Seyedehshahrzad Seyfafjehi[*]

Professor Neutron: Sixty percent! Professor Sun, don't you think living off of solar energy on Mars which has only sixty percent sunlight intensity of Earth sounds like... I'm looking for the right word... Madness?!

Professor Sun: Professor Neutron, I do understand it sounds inefficient thinking about the long term. But for initial settlement on Mars it is a viable option! After all nuclear reactors are also a solution for a decayed or so.

Professor Neutron: Nuclear reactors are sustainable, just send a balloon to Jupiter's atmosphere, harvest helium 3 as much as you need, viola! Your reactors are refueled.

Professor Sun: And what about the nuclear waste? Would you go viola! And get disposed of them?! I highly doubt it professor, I mean you know better than me how hazardous the residue of nuclear waste is and yet we haven't found a solution for getting disposed of them here on Earth... this is while solar energy is not only environmentally friendly...

Professor Neutron: Ms. Sun! for some reason you keep forgetting about sand storms on Mars that block the sunlight for even weeks!

Professor Sun: Mr. Neutron, do I need to update you on technology? even today we have solar batteries lasting for if not months, for weeks. Think about NASA's space probe using solar power that reached Jupiter. I repeat Jupiter which is even farther away than Mars!

Professor Neutron: So you are comparing energy needed for only a space probe to energy needed for a whole human colony...!?

Professor Steam: Colleagues! Don't you see the necessity of geothermal energy now?! It's environmentally friendly, no hazardous residues or waste, it is renewable and always usable, no storm to obscure it. With volcanos we have found on Mars, Olympus Mons with twice the height of Mount Everest's height ready to erupt, we definitely know that Mars has thermal heat. And yes Mars' core is cooler than earth but if we drill deep enough and create the Geothermal power plant we need, accessible hot fluid and an overlaying accessible cold fluid in order to be used as heat sink, to overcome the problem of heat rejection on Mars since there is no atmosphere we can happily use geothermal energy for our human colonies on Mars!

Everyone stared at Professor Steam with a solemn expression. Smell of sweat and burned out cigarettes filled the room. A fly was buzzing..

---

[*] Seyedehshahrzad Seyfafjehi, Department of Computer Science, Bilkent University,. Ankara/Turkey

Moving to Mars
Staying Warm on Mars

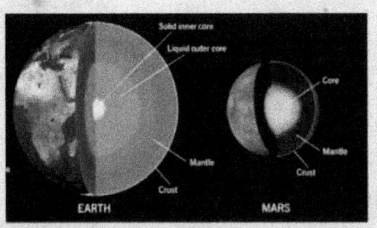

 Solar Energy:

 Geothermal:

 Nuclear Power:

We Evolve to Build the Future
Energy for life

	Energy Resources	Earth vs Mars
☀	>Environmentally friendly >Renewable >Inexpensive tech	>Insufficient for Mars' condition >Dust Storms Blocking the sunlight for weeks >Sunlight intensity 60% of the Earth's
♨	>Viable Renewable Existance of volcanos on Mars proof of thermal heat >Environmentally friendly	>High tech: >Deep drilling >No atmosphere: heat escape >Cooler crust and core than the Earth.
☢	>Sustainable >Carbon free production	>Nonrenewable >Nuclear waste disposal : >leaves hazardous residues

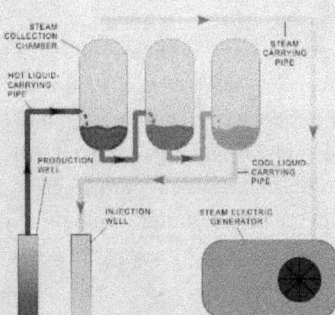

Geothermal power plant on Mars

Power plant structure:
Accessible hot fluid and an overlaying accessible cold fluid in order to be used as heat sink.

Challange:
The mean surface temperature on Mars is about -63°C and the mean pressure is about 7mb, this is while the mean surface temperature on Earth is 15°C and the mean pressure is about 1013mb.

Result:
Cooling in energy extraction from geothermal fluids prevents liquid water from being stable at the surface.

SHOP WITH A GRAIN OF SALT

Büşra Yücel[*]

Presumably nobody wants to eat plastic, but today everyone unknowingly does, thanks to the growing consumption of plastic bags and it's time for the consumers to put an end to this with a simple, sustainable switch to reusable bags that they can actually make at home. The pollution that plastic bags create in the oceans are not limited to only seafood, but to salt as well, as plastic pollution doesn't completely degrade over time but breaks down into micro plastic, tiny grains of plastic that cannot be separated from salt. So even if one avoids seafood not to include plastic in their diet, the fact that plastic is now integrated in salt makes it very difficult to escape eating plastic. The dark sides of these innocent, free plastic grocery bags also extend from killing marine animals in increasingly large numbers every year, to wasting a billion litres of oil, 8% of the world's existing oil, and costing billions of dollars every year to produce and additional billions of dollars to clean up afterwards. Stopping this is only possible with the consumer, making the simple change of switching to environment friendly, reusable cotton bags instead of energy wasting, harmful and actually pricey single use plastic bags while grocery shopping can save the planet. The good news is you don't even have to go out and buy these reusable, eco-friendly, cotton bags, you can make it at home by recycling your old t-shirts without little to no effort which makes them not only eco-friendly but also washable, practical, free, long-lasting, creative and custom. Even if cotton also uses up a lot of water resources, recycling them eliminates this problem as well. All you need is some scissors, an old t-shirt and a little imagination to save the world. By grocery shopping responsibly, we might start living in a world where we don't have to eat plastic with our salty meals anymore, marine animals live a happy, long life and we save our energy resources and money. The world depends on our old t-shirts and imagination right now, so it's time for everyone to become a hero while grocery shopping!

---

[*] Büşra Yücel, Department of Psychology, Bilkent University, Ankara/Turkey

SHOP WITH FABRIC, STOP EATING PLASTIC

WHY SHOULD YOU SAY GOODBYE TO PLASTIC BAGS?

- One plastic bag is used only for 12 minutes, but it takes more than 500 years for a single plastic bag to degrade.
- Except, they don't entirely degrade but turn into microplastic.

- The microplastics in the oceans are now in our **salt**, our food.
- Annually, we pollute the oceans with 81 billion kg of plastic.
- 100,000 marine animals are killed by plastic bags every year.

- 16 billion liters of oil is used to make 100 billion plastic bags.
- It only takes 14 plastic bags to drive your car for 16 km.
- 8% of the world's oil is used in the production of plastics.

- If Turkey banned plastic bags, theoretically, 6.5 billion MJ energy could be saved, greenhouse emissions could be reduced by 395 ton and solid waste by 16 million kg.

- The USA spends 4 billion dollars annually for plastic bags.
- California spends 25 million dollars to clean plastic bags.
- Thousands of dollars are spent on landfill plastic strategies.

WHICH MATERIAL IS BETTER AS AN ALTERNATIVE?

MATERIAL / WASTE	ENERGY CONSUMPTION	GREENHOUSE EMISSION	LITTER	WATER USAGE
SINGLE-USE PLASTIC BAG	⚡⚡	🌫🌫🌫	🗑	💧
BOUTIQUE PLASTIC BAG	⚡⚡⚡⚡⚡	🌫🌫🌫🌫🌫🌫	🗑🗑🗑🗑	💧💧💧💧
PAPER BAG	⚡⚡⚡⚡⚡⚡⚡	🌫🌫🌫🌫	🗑🗑🗑🗑🗑	💧💧💧💧
REUSABLE COTTON BAG	⚡	🌫	🗑	💧💧💧💧

MAKE YOUR OWN CUSTOM, REUSABLE, ECO-FRIENDLY GROCERY BAG:

1. GET AN OLD T-SHIRT

- Just make sure that it's cotton!
- The bigger the t-shirt, the bigger the bag!

2. START CUTTING

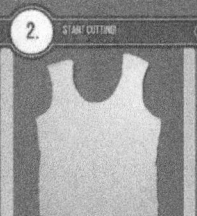

- Cut the sleeves, collar and the bottom of the t-shirt.
- You don't have to be precise!

3. CUT SMALL STRINGS ON THE BOTTOM

Cut the bottom of the t-shirt into approximately 2x6 cm parts (no precision necessary).

4. TIE THE STRINGS TOGETHER

- Double-knot the front pieces to their back pieces one by one.
- Cut fringes for a cleaner look.

5. TIE THE EXTRA PIECE

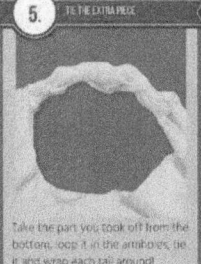

Take the part you took off from the bottom, loop it in the armholes, tie it and wrap each tail around!

6. DECORATE IT AS YOU WISH!

- Don't like fringe? Turn it inside out!
- Customize your one of a kind, washable, reusable grocery bag!

Büşra Yücel 21502645

CLIMATE CHANGE IS ON OUR PLATE
Ezgi Altınöz[*]

The meat on our plates are destroying the planet! Considering the destruction of the planet by our own hands caused by our consumption of reified "foods" that come in the shapes of a hamburger or a meatball, the vegan way of life that is considered to be as a challenging way of life-style nowadays, is the saviour at this point. Our habitat consumption increases with wildly growing meat, milk and all animal derived food consumption and we do not even think of our planet's future. The livestock factor is the leading cause in desertification of one-third of the world. And it is the killer of the Amazon forests, which are holding about 120 billion tons of carbon dioxide. Animal feed production and cattle ranching, which caused 91% of the Amazon rainforest's desertification is therefore the leading cause of climate change.

Furthermore, the emission of methane gas is 20 times more damaging than carbon dioxide emissions. 37% of the methane emissions are from the livestock industry. Considering that a cow emits methane gas of 70 to 120 kilograms per year, is it possible to imagine the effects of methane gas released by 1.5 billion cows during their lifetime? The growth of the livestock industry with increasing meat consumption destroys our planet. Meat consumption exceeds the emissions from all transportation-caused emissions with a ratio of 18% in greenhouse gas emissions. Stop this! If you go vegan, you will not be contributing this massacre of the environment. It's time to refresh our plates and save our planet!

---

[*] Ezgi Altınöz, Department of Communication and Design, Bilkent University, Ankara/Turkey

CLIMATE CHANGE
IS ON OUR PLATE

Footprints by diet type

Greenhouse Gas Emissions from Proteins and Vegetables

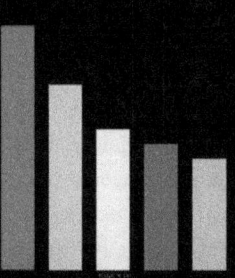
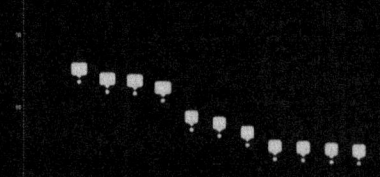

DEFORESTATION

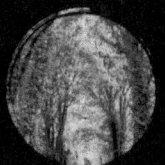

The Amazon stores roughly 120 billions of tons of carbon.

Forests are a crucial carbon stock: forest ecosystems globally store about one-and- a- half times as much carbon as is present in the atmosphere. They act as a brake on further acceleration of climate change.

GHG EMISSIONS

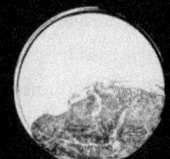

Around 18% of global emissions of greenhouse gas are related to meat consumption worldwide.

A beef calf produces five tones of CO2 - equivalent throughout its life cycle.

FARTING COWS

Belched methane from livestock, plus methane from manure make up 37 % of global methane emissions. Methane is 20 times as damaging to the climate as CO2.

Pursuing a vegan diet will reduce one's CO2 footprint more than half comraped to a meat lover.

SOURCES:
Meat Eater's Guide by Environmental Working Group
Amazon Cattle Footprint by Greenpeace

Prepared by Ezgi Altınöz

CARBON CAPTURE STORAGE: DECREASE EMISSIONS – INCREASE EFFICIENCY

Naz Alara Erbek*

Greenhouse gas emissions resulting from the use of fossil fuels to generate energy mostly cause the carbon and carbon dioxide gases to be released into the atmosphere and this leads to global warming and climate change. In 1997, it was decided in the Kyto Protocol that the distribution of CO_2 would fall below 1990 level. To prevent the global warming and climate change, experts are searching for new methods and ideas. Carbon Capture and Storage (CCS) is one them. Basically, CCS system captures CO_2 from electricity generation and industrial production. CO_2 is liquefied with high pressure and stored between special impermeable layers of the ground. There are various techniques of CCS.

For example, CO_2 can be transported by pipes or tankers. CO_2 can be added incertain places to improving the production efficiency in oil wells and increasing the production efficiency in natural gas wells. With this method, the quality of the oil and gas can be increased.

CO_2 can be stored in deep coal vessels, deep salt formations or oil and gas reserves. CCS is mainly used in developed countries. In Turkey there are certain common beliefs about CCS system. According to researches, %46 of the people think, CCS could not be used in short-medium term in Turkey. However, according to same research, it shows that, only %51 of these people know the system of the CCS. Accordingly, people should get access to the information about this method, to protect the environment and prevent the global warming.

---

*Naz Alara Erbek, Department of Communication and Design, Bilkent University, Ankara/Turkey

CARBON CAPTURE STORAGE

Greenhouse gas emissions **resulting from the use of fossil fuels to generate energy mostly cause the carbon and carbon dioxide gases to be released into the atmosphere and this leads to** global warming **and** climate change.

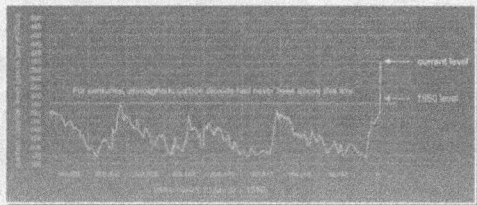

NASA, Global Climate Change

According to researches, greenhouse gas emissions increased dramatically over the years.

Many scientists share the same view that the CO_2 **emissions in the world** should be reduced by at least 50% compared to today.

According to the reports of the International Energy Agency, **Carbon Capture Storage (CCS) has the** potential to contribute 14% **to global carbon emission reduction based on 2060.**

With this rate, CCS is the third critical factor behind Energy Efficiency (40%) **and** Renewable Energy (35%).

A LETTER TO OBLIVIOUS HUMANS
Sardar Talal Khalid[*]

Dear oblivious humans,
If I could go to the police station to file an official FIR against more than the 50% of the world's population (who are oblivious to global warming) on the charge of being responsible of potentially killing or harming my future grand children in a couple of decades, I would do so. But sadly no one would take me seriously and I would be back to square one.

Now I am not as radical a person to say that global warming is a collective suicide or collective mass murder. That would be ridiculous. But seriously, what are we thinking? Do we as a society not realize that global warming is NOT a pending threat but it is an immediate threat as pointed out by almost 97% of scientists? I am highly fascinated by a civilization advance enough to land on the moon but psychologically impaired to choose between preventing global catastrophes and its urges to use environmentally unfriendly products for instant gratification.

The industrial revolution has led to such an increase in the amount of toxins in the environment that the planet is shaking. It can't even bear the high amount of Carbon Dioxide constantly being pumped into the air anymore (which previously was an essential component of air). This realization is dawning upon most of us, but the effects have already taken impact.

This is to sincerely belittle humans. I understand that in the digital age we have a plethora of information on the internet mixed with misinformation and conspiracy theories which spreads misconceptions. I also understand that psychological distancing causes people to forget about the threats which can occur in the future. But still. Certainly, there is a long way before we convince people that killing the planet is not a good idea. What irks me, however, is that there is in fact a need to convince people to not kill the planet.

Sincerely,
Concerned humans

---

[*] Sardar Talal Khalid, Department of Communication and Design, Bilkent University, Ankara/Turkey

NO MORE GASOLINE CARS? THE WORLD IS TURNING TO ALTERNATIVE FUELS - TURKEY IS NOT FALLING BEHIND

Feyza Yılmaz[*]

In the past year, the use of hybrid vehicles in the world has increased by 54 percent and the use of electric vehicles has increased by 62 percent. Which may be the omen of a heart breaking break up with diesel and gasoline cars, who knows?

The reason why these vehicles are being preferred is not only they are environmentally friendly in terms of fewer gas emissions, but also they offer many additional benefits. Hybrid and electric cars have less fuel cost- average fuel cost of a gas powered vehicle is almost two times of hybrid cars and three times of an electric car-, more reliable with less maintenance cost and higher resale value. So, if we want to save some money and to avoid a possible doom scenario due to climate change, in addition to planting trees we may start using electric cars.

Turkey is not oblivious to the raise of alternative fuels usage in the World. Country is having a rapid change in usage of alternative fuels. In the past year, the hybrid vehicle market increased by 370 percent, with the increase of 950 sales to 4451 sales in a year. Besides, it is visible that tendency to electric cars increases by years. There are 573 electric cars in the streets around the middle of 2018 while the number of electric vehicle stations reached the number 1500. In addition, there are 50 electric buses that are in use in Turkey which will reach the number of 250 in coming year. Judging by this increase, Turkey cannot remain indifferent to the "go green" movement and it can be seen that hybrid and electric vehicles will be so much more popular in coming years. You go Turkey!

---

[*] Feyza Yılmaz, Faculty of Business Administration, Bilkent University, Ankara/Turkey

RISE OF ELECTRIC AND HYBRID CARS IN TURKEY

Reasons To Disseminate Electric and Hybrid Cars

- Reduced fuel costs (electric vehicle costs more than 6 times of gas vehicle)
- Less pollution due to fewer emissions, way of a greener future
- Especially electric cars require less maintenance, eliminates maintenance cost
- Instant torque
- No idling
- Higher resale value
- Electric cars are safer than gasoline cars because they do not catch fire in accidents

Sales of Hybrid & Electric Cars are Increasing

Increase in total sales of hybrid and electric cars is almost %370. Market share and availability are rising rapidly.

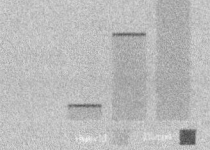

Annual Emission of Conventional Gas is Quite High in Comparison

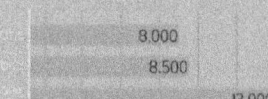
- 8.000
- 8.500
- 13.000

Electric Car Sales in Turkey are Changing Rapidly

It is visible that tendency to electric cars increases by years. Usages of alternative fuel sources in cars – such as electric– are not that common in Turkey, but it has progress potential.

Compared to Gasoline Cars Lifetime Consumption and Fuel Costs of Hybrid and Electric Cars are Quite Low

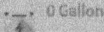 0 Gallons
$5,200

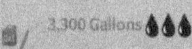 3,300 Gallons
$9,600

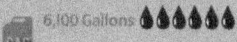 6,100 Gallons
$18,000

Ratio of Electric Cars are Still Extremely Low in Turkey

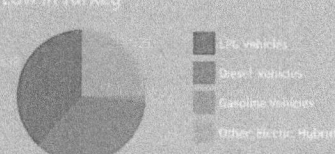
- LPG Vehicles
- Diesel Vehicles
- Gasoline Vehicles
- Other: Electric, Hybrid

Fuel Stations of Electric Cars Reached To A Convenient Level

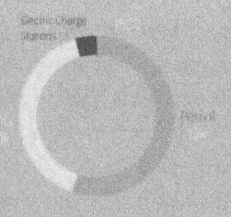
- Electric Charge Stations
- Petrol

Reference: TEHAD- Turkey Electric & Hybrid Cars Association

FEYZA YILMAZ

HOW THE WORLD AND TURKEY SEES ELECTRICAL VEHICLES
Onurcan Boran*

Electric Vehicles (EV) are more advantageous than internal combustion engines (ICE) because they run on electricity and electricity comes from many sources like solar, wind and hydropower. However, petrol isn't available everywhere and it is expensive. Countries like USA, Western Europe, and Eastern Asia are trying hard to change their transportation policies. Norway aims to completely electrify the transportation until 2025 and Norway abolished import tax of electrical cars and imposed high taxes on ICE. Since 1978 US made strict laws to decrease CO_2 emissions from cars. In Japan there were exactly 3,300 electrical cars on traffic. Japanese government created a project called "the Clean Energy Vehicles Introduction Project". This project was aiming to provide citizens with the half of the price of electric cars. Although China has many weaknesses in EV policy compared to Japan China produces cheapest car batteries. In Turkey, we still don't have a local and national electrical car and charging stations are only common in western part of Turkey. East of Turkey seriously lacks charging stations. More importantly, national income and level of welfare is not sufficient to purchase electric cars (EC). Lastly, Turkey had better to make use of EVs because Turkey is rich in renewable energies.

As for charging stations, they are really advantageous. For example, in electrical car models, they have an interrelated navigation system. Whenever car runs out of energy, it changes its behaviour and tries to go to charging station.

But there are still problems! Despite all advantages, there are still some problems with charging stations in general like technological, economic and social problems. Electrical cars are more expensive than internal combustion cars. In ordinary charging stations it takes 8 hours to charge. Even in developed countries like US, people still aren't so familiar with charging stations. According to a survey in US, 29 percent of citizens say there are too few, if any, public charging stations where I travel.

---

* Onurcan Boran, Department of Communication and Design, Bilkent University, Ankara/Turkey

CAN PLUG-IN ELECTRICAL VEHICLES COMPLETELY ELECTRIFY DEVELOPED REGIONS IN THE WORLD?

Outstanding Environmentalist features of Plug-in Electric Cars:
1. Emits no tailpipe emissions.
2. Production of their batteries does not have environmental impact. These batteries contains lithium and there is an abundance of lithium reserves. We have sufficient reserves until 2050. Also, most of these batteries are recyclable.

ELETRICAL VEHICLE CHARGING STATION MAPS OF DEVELOPED REGIONS

ASIA = 1,130 Supercharger Stations with 8,496 Superchargers

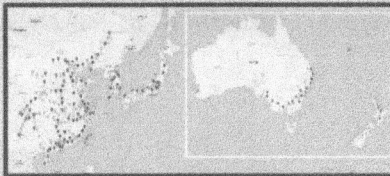

WESTERN EUROPE = 1,130 Supercharger Stations with 8,496 Superchargers

- Developed countries has increased number of public charging stations. This shows that developed countries like USA China, Australia and Western European countries feel responsible for air pollution and they are really trying hard to stop air pollution.
- Turkey is a developing country and industry is developing fast. Turkey needs to take some steps about electric vehicles and supercharger stations.
- East of Turkey lacks charging stations.

- Sales of electrical cars still are not so high because national income and level of welfare is not sufficent to purchase them.
- Actually, Turkey should make use of EVs because Turkey is rich in renewable energy resources.
- In some electrical car models, they have an interrelated navigation; Whenever car runs out of energy, it goes closest charging station.
- Although this technology has many benefits, all societies including Turkish society is still unaware of it.

USA = 1,130 Supercharger Stations with 8,496 Superchargers

TURKEY = 1500 Electric Charge Stations / 13,000 Petrol Stations/ 10,000 LPG Stations/ 9 Superchargers

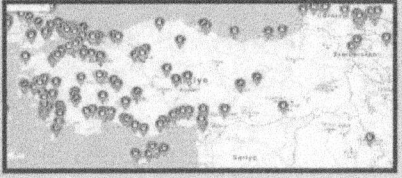

DATA REFERENCES

- Farhan, Mohd. Marketing of Electric Cars. Berlin: Technische Universität Berlin, 2016. Print p. 30-54
- "PlugShare - Find Electric Vehicle Charging Locations Near You" www.plugshare.com/.
- "Supercharger | Tesla." Tesla, Inc. www.tesla.com/supercharger.
- "Electrical Vehicle Survey Methodology and Assumptions." 2016.

Onurcan Boran

ABOUT THE AUTHORS

Nursan AKINCI was born in 1996 in Ankara Turkey. She took her elementary education at İlkem Private School. After, she graduated from Yüce Private High School in 2014. She studied at Bilkent University Faculty of Fine Arts, Design, and Architecture, Communication and Design Department because of her interest to design and art. She graduated from Communication and Design in 2019. She did her first internship in Publicis One İstanbul, Starcom Mediavest Group in the field of media planning. She completed her second internship in Dwt Mandalina Ankara which is an advertising agency. In Dwt Mandalina, she worked in the department of strategic planning and copywriting. In upcoming period, she is planning to focus on advertising and psychology.

Ezgi ALTINÖZ was born in 1998 in Balıkesir, Turkey. She graduated from Salih Korkut Budaras Teacher Training Anatolian High School in 2016. She studies at Bilkent University Department of English Language and Literature and she does her minor at Bilkent University Faculty of Communication and Design. She completed her internship in Hürriyet Ankara as a journalist. She worked for local news based in Ankara. She is now an exchange student (ERASMUS) at TU Dortmund University.

Turan BAYRAM was born in 1995 in Kayseri, Turkey. He took his elementary education in Milli Zafer Primary School. Later, he graduated from Gökçeada Atatürk Teacher Training High School in 2013. He studies Bilkent University Faculty of Communication and Design Department and will graduate in 2019. He did one of his internship in Ankara, and in summer 2019, he will attend the second one. His first internship was with Nöbetçi Ajans, a small documentary and advertisement company. In Nöbetçi Ajans, he mostly worked on the editing and writing. He is currently working on a project with World Health Organization (WHO) called Refugee Health-oriented Journalism. His main interests are history, journalism and new media, and plans to work on these in the future

Onurcan BORAN was born in 1995 in Kars, Turkey. He took her elementary education at Kars Atatürk Primary School. Later, she graduated from Kars Cumhuriyet High School in 2013. He studied at Bilkent University Faculty of Communication and Design Department and graduated in 2020.He did his internships in Ankara. He completed her first internship in Kampüs News Agency. For the second internship, he worked in Bereket TV as a reporter. In Kampüs News Agency, he worked for writing online news. In Bereket TV, he worked for social media accounts of clients. She also worked for writing bulletins. In upcoming period, he is planning to focus on media and film studies in academy.

Orhun Ege CANSARAN was born in 1996 in Ankara, Turkey. He took her elementary education at many schools from many part of Turkey. Later, he graduated from Rize Teacher Training High School in 2014. he studied at Bilkent University Faculty of Communication and Design Department and graduated in 2019.he did her internships in Ankara. he completed her first internship in Haberturk Ankara. For the second internship, he worked in Konya Chamber of Commerce as a PR person and magazine editor. In Haberturk, he worked for writing news and making interviews. In KTO, he worked for public relations and preparing the official magazine of the chamber. In upcoming period, he is planning to change main discipline of him and get a master degree from politics and international relations to be more efficient in his future life.

Ruben DÜCHTING is Dipl.-Journalist and PR Consultant. He studied Information Technology at University of Cologne and Technique Journalism University of Applied Science Bonn-Rhein-Sieg. Since 2008 at iserundschmidt GmbH – agence for science communication in the section of text, concept and PR consultancy.

Naz Alara ERBEK was born in 1997 in Bursa, Turkey. She took her elementary education at Bahçeşehir College. Later, she graduated from Büke High School in 2015. She studied at Bilkent University department of International Relations and she will graduate in 2020. She studied Communication and Design as a minor programme. She did her Erasmus Exchange Programme in Masaryk

University, Brno, Czech Republic. She did her internship in Regional Environment Center (REC) Turkey in Ankara. During her internship she prepared reports, researched and collected data, supporting on project development about climate change, environmental policies and sustainability. In upcoming period, she is planning to focus on project management, urban governance and development.

Rustam GULIYEV was born in 1995 in Baku, Azerbaijan. At the age of 7, he became the laureate of the contest "On the Wings of Fantasy" organized by British Airways. In 2011 he published his first fantasy book both in French and Russian, later in 2013 the exhibition dedicated to 'art of dinosaurs' was held in France. He graduated from 23d school named after Alexander Pushkin in Baku. He studied at Bilkent University Faculty of Communication and Design Department and will be graduated in summer 2019. In upcoming period, he is planning to focus on character design, animation and concept art.

Prof. Dr. Seldağ GÜNEŞ PESCHKE was graduated from Ankara University Faculty of Law. She finished her master and phd in Ankara University Institute of Social Sciences. In 1995 she started to work in Privatization Administration as a lawyer. She has started her academic career in Gazi University Faculty of Law as research assistant in 1997. By the scholarship of Italian Government between 2000-2001, she did research for her phd. in Roma La Sapienza University.
She got scholarships from DAAD and Frankfurt Max Planck Institute several times and worked in different German Universities between 2006-2017. In 2009 she became associate professor in Gazi University Faculty of Law. Since February 2015 she is working as a professor of Comparative Law in Ankara Yıldırım Beyazıt University Faculty of Law. She has taken part in EU projects on refugees, women issues, research integrity. She speaks fluent English, German and Italian. She has five books and many articles in national and international journals. She works mainly on comparative law, family law, women studies, social media and media law, personality rights, ethics, data protection, contracts law, children rights and Roman Law.

Sardar Talal KHALID was born in 1996 Islamabad, Pakistan. He took his elementary education at Roots International College. Later, he studied in the Communication and Design Department in Bilkent University graduated from

Bilkent University. He extensively interned in Pakistan in PTV and Bahria TV broadcasts as a planner and writer for shows, and later interned in Barcelona as a communicator and promoter for Uniplaces. He has delved into film and documentary productions and continues to produce more in the future as his career.

Öykü ÖNCÜL was born in 1995 in İstanbul, Turkey. She took her elementary education at Feyzullah Turgay Ciner Elementary School. Later, she graduated from İstek Private Semiha Şakir High School in 2014. She studied at Bilkent University Faculty of Communication and Design Department and graduated in 2019. She did her both voluntary and mandatory internships at İstanbul, Ankara and Berlin. She completed her first voluntary internship in Hürriyet Newspaper İstanbul. For the second internship, she worked in L'officiel Magazine as an editor and translator. In Hurriyet, she worked at Agenda Section and responsible for writing news and selection of the image's. For the third internship, she worked in Güzel Sanatlar Advertisement Agency in both Strategical Planning and Creative Departments of the agency. In Güzel Sanatlar Advertisement Agency she worked for copywriting and strategical planning of both traditional and digital media advertisements of several prominent brands of the Turkey. She completed her third internship at DWT Mandalina Agency Ankara where she worked as a Social Media Copywriter. For the fourth internship, she worked in Vitamin Marketing Agency at İstanbul. In Vitamin Marketing, she worked for social media accounts for clients and sports event planning. For the final internship, she worked in Axel Springer Digital News Media Company. In Axel Springer, she worked for vertical product management, partner integrations, and digital marketing development. In upcoming period, she is planning to focus on Digital Media Marketing.

M. Mert ÖRSLER was born in 1995 in Eskişehir. He took her elementary education at Çankaya Elementary School in Ankara. Later, Mert graduated from Eskişehir Gazi Mustafa Kemal Anatolian High School in 2013. Receiving a full academic scholarship, he studied at Bilkent University Department of Communication and Design and graduated as valedictorian in June, 2019. Mert did his internships in Ankara. He first completed an internship in ATV Ankara. For the second internship, Mert worked in TRT Headquarters and Ankara Studios as a new media intern. In ATV, he was working for the news desk and responsible for tasks related to photojournalism as well as news writing, editing and design. In TRT, Mert edited videos, designed graphic-based works, contributed to webs design practices and experienced online broadcasting. In upcoming period, he is planning to continue his academic life and research in Canada in the area of communication and media studies.

Dr. Dr. Lutz PESCHKE born in 1964, studied Chemistry (Ph.D at University of Heidelberg/Germany) and Media Studies (Ph.D University of Bonn/Germany). Since 1999 head of the Department for Multimedia in iserundschmidt GmbH - Agency for Science Communication. From 2010-2012 lecturer for public relations and science communication in the Department of Media Studies at University Bonn. 2013-2014 lecturer for Media Design in the Faculty of Architecture of Çankaya University in Ankara. Since 2015 lecturer, since 2018 assistant professor for Media and Communication Studies in the Department of Communication and Design at Bilkent University in Ankara. Research interest: science communication, new media studies.

Simge SADAK was born in 1996 in Ankara, Turkey. She took her elementary education at Bilkent Elementary School. Later, she graduated from Bilkent High School in 2014. She studied at Bilkent University Faculty of Communication and Design Department and graduated in 2019.She did her internships in Istanbul. She completed her first internship in ELLE Magazine, Istanbul. For the second internship, she worked in TV8. In ELLE Magazine, she worked for writing posts and editing. In TV8, she worked for interior productions and backstage. In upcoming period, she is planning to focus on editorship and advertising.

Özgün Evrim SAYILKAN was born in 1995 in Adana, Turkey. She took her elementary and high school education in Adana. Later, she moved to Ankara in 2013. She studied Communication and Design Department at Bilkent University Faculty of Art, Design and Architecture and graduated in 2019. She did her internships in Ankara. She completed her first internship in Cosmic Creative, a full-service advertising agency. For the second internship, she worked in You Media Magazine as a content creator. During her college education, she worked at Institutional Relations and History Unit of Bilkent University as an operator. In upcoming period, she is planning to focus on media and cultural studies.

Seyedehshahrzad SEYFAFJEHİ was born in 1996 in Tehran, Iran. She took her elementary education at Raherosh school. Later, she graduated from 22 Bahman Iranians' school in Ankara in 2014. She studies at Bilkent University Faculty of Computer Engineering department and will graduate in 2019. She did her internships in PrimeTek as a web developer. In Supply Chain Wizard, she worked for quality testing and assurance. In ICL International College of Languages, she worked at the advertising department. She also worked for Bilkent news writing news stories. In upcoming period, she is planning to pursue a master's degree in Communication, Media and Journalism.

Manfred SCHMIDT studied philosophy, German literature and history in Cologne and Dusseldorf and is docent author and creative director with several publications. CEO of iserundschmidt GmbH – agency for science communication in Bonn/Germany. He is responsible for concept, strategical and media consultancy.

Marleen-Christin SCHWALM is Art Directior Graphics at iserundschmidt GmbH. She completed her apprenticeship as media designer at iserundschmidt between 2010 and 2012 and absolved her advanced training for Industrial Management Assistant Digital and Print Media.

Feyza YILMAZ was born in 1997 in İzmir, Turkey. She took her elementary and high school education at İzmir. She studied at Bilkent University Faculty of Business Administration with an honor degree and graduated in 2019. During her university years, she finished her Erasmus program in Universitat de Barcelona. Also, she worked with several university clubs as member and organizer of several events- YES,BT-, she volunteered for Lösev, TDP and Make a wish foundation, She delegated for her university in several Estiem events and participated in Erasmus plus projects. She did her internships in istanbul. She completed her internship in Zebramo as a marketing project intern. In her internship she planned, scheduled, created and published advertisements and creative content for Zebramo. In upcoming period, she is planning to focus on marketing and marketing communications.

Büşra YÜCEL was born in 1996 in Ankara, Turkey. She took her primary education at Ankara Yüce College and went to middle school at Ankara Nesibe Aydın Schools. She later graduated from Ankara Yüce College High School in

2015. She graduated from Bilkent University Psychology Department in 2019. She completed her internship at Renk Family Counseling in Ankara as a clinical psychology assistant. In the upcoming years, she plans on developing her design skills alongside her career in psychology.

www.ingramcontent.com/pod-product-compliance
Lightning Source LLC
Chambersburg PA
CBHW050240230526
45470CB00005B/2044